Dedication

*This book is dedicated to every soldier
who ever found humor in hardship, and
to those who cooked, fought, and
   laughed their way through hell — because
someone had to feed the fight.*

Chapter Outline

Chapter One – *Eighteen and Bulletproof*

Chapter Two – *Cooking Under Fire*

Chapter Three – *Shit Burning and Steak Turning*

Chapter Four – *Midnight Chow and Incoming Mortars*

Chapter Five – *Convoy Cooks*

Chapter Six – *Green Zone Mirage*

Chapter Seven – *Mail Call and Memories*

Chapter Eight – *Sandstorm Sundays*

Chapter Nine – *The Burn Pit Baptism*

Chapter Ten – *The Shit Talk Olympics*

Chapter Eleven – *Patrol Paratroopers*

Chapter Twelve – *Six More Months of Sand and Sanity*

Chapter One: Eighteen and Bulletproof

I was eighteen, cocky as hell, fresh out of high school, and thought I had life figured out. I'd just enrolled at the local community college in Muskegon, Michigan. My plan? Blow my book stipend on Columbia House — ten CDs for a buck, no intention of ever paying — and spend the rest on beer for the weekend at Out of Bounds Night Club.

It was the early 2000s, everyone from every local high school crammed together, and I figured college would be one long party. It lasted about three weeks before I realized I needed to get the hell out of town before I turned into everyone else who said they were "going to leave someday."

One of my best childhood friends, the kid who lived across the street, told me he was talking to an Army recruiter. That planted the seed. I called up my high school best friend, and within a week we had a plan: join the Buddy Program, ship out together, go through Basic, AIT,
Airborne School, and get stationed at Fort Bragg — home of the 82nd Airborne Division.

I thought joining the Army would be like the commercials — slow-motion jumps, brotherhood, thunderous music. They never showed the part where you're elbow-deep in scrambled eggs that came out of a box labeled "Dehydrated — Do Not Inhale, Unless You Want Your Lungs Destroyed Like walking into a cloud of Mustard Gas."

By the time my high school friends were faking IDs to get into bars, I was wheels-up from Fort Bragg, headed overseas. I'd soon be standing on a tarmac in Kuwait City in March of 2003, sweating through my uniform and trying not to melt. The desert heat didn't just sit on you — it crawled inside your bones. And I was about to help feed a few thousand paratroopers and support units who were unknowingly weeks away from jumping into Iraq for the initial invasion.

Not exactly the cooking job my dad had in mind when he taught me how to make marinara sauce at Sorrento's Restaurant back in Muskegon. He'd slide a pizza pan across the counter sauced up with fresh tomato sauce and say, "Son, cooking is patience — don't rush the flavor." Patience wasn't exactly the Army's strong suit.

The first time I opened a military field recipe book — TM 10-412, Army Recipes — it felt more like assembling ammunition than food. Everything came in bulk: 50-pound bags of flour, industrial tubs of peanut butter, and instructions like "yields 100 portions." The Army didn't cook; it mass-produced morale. Mind-fucks in bulk.

March 23rd, 2003, the night before my birthday, there was a chill in the air, and it was time to hit the cot in the tent we had built from the ground up. My comrades — a bunch of lovable assholes — thought it'd be funny to fill my entire sleeping bag with baby powder and then douse me with water the moment I climbed in. "Happy birthday, Peltier!" they yelled, like a powder-covered baptism into manhood.

My birthday, the 24th of March, that morning in Kuwait, we were up before dawn. The sky had that ghostly blue tint that only appears right before the sun decides to fry the horizon. The chow tent smelled like burnt coffee, helicopter smoke, and men who hadn't seen a shower in days.

"Hey, Peltier," one of the sergeants barked, "you better make those eggs taste like Waffle House or I'm going to request KFC — Kuwait Fried Camel."

"Roger that, Sergeant," I said, trying to sound confident while stirring what looked like yellow cement. The secret? A little salt, a touch of pepper, and a prayer that nobody noticed. Because Lord knows if you don't annotate on your recipe card that you put a little bit of love in your food, you were going to get the shit smoked out of you.

We didn't have real pans — just massive steel griddles that hissed like snakes when the powdered eggs hit them. The 82nd Airborne was preparing for war, and here I was making breakfast for heroes. It felt strange — too normal for what was coming. One guy walked by, M4 slung over his shoulder, grabbed a tray, and said, "You know, Peltier, you're the real MVP. If the eggs are good, we'll fight harder." He grinned, and I realized food was more than food — it was sanity.

I don't think anyone really knew what was getting ready to unfold. We were in Kuwait for training — for the what if, the just in case, right? My Father's Dad had spent a lifetime and was a full bird colonel in the Army. He used to tell me stories about leadership, discipline, and honor. But I don't think he ever told me how to make five hundred servings of eggs in a tent during a sandstorm. The heat coming off the ovens chafing your nuts and singing your eyebrows.

My mother's father — my Opa — was a tailor from the old Yugoslavia, captured in World War II and held as a prisoner of war for years before escaping. He came to America with needle scars on his hands and silence in his eyes. He passed in 1976, long before I was born, but his story was the thread that ran through my family. My mother came to the U.S. as a baby, sponsored by a church that believed in second chances. Maybe that's why I joined the Army — the idea that service could stitch broken things back together.

I grew up in a blended household, a puzzle with too many missing pieces — six older sisters and me, the baby boy. Two were step-sisters from the man my mom married when I was five.

To this day, I swear he was sent straight from Hitler's leftovers.

That morning in Kuwait, we were on Failaka Island doing mount training. I stood there flipping eggs with a metal spatula, surrounded by soldiers laughing about home, women, and beer they couldn't have. The radio crackled. The war was about to start. And I remember thinking — this is it. This is where the boy from Muskegon becomes a man.

It wasn't the mortars, or the RPGs cutting the sky open, that changed me. It was the laughter over Rip-Its, the friendships forged between bites of powdered eggs, and the realization that even in hell, you can still feed hope.

Army Recipe — SCRAMBLED EGGS,
DEHYDRATED (FIELD VERSION)
Source: TM 10-412
Yield: 100 portions (or one squad of
very   hungry   paratroopers)
Ingredients:
- 5 lbs dehydrated egg mix
- 4 gallons water
- 2 lbs butter or oil substitute
- Salt and pepper to taste
- Optional: Tabasco bottle saved from an MRE, courage,
  and a sense of humor.
  Directions:
1. Reconstitute egg mix with water until smooth.
2. Heat griddle to "sizzling." Add oil.
3. Pour eggs and stir constantly with a long-handled
   spoon to prevent scorching. (Because everyone in the
   military has eaten scorched eggs.)
4. Serve hot. (Add jokes liberally to taste.)
Sergeant P's Note: If it's edible, I made it. If it tastes like
shit, someone else made it.

Chapter Two: Cooking Under Fire

Every soldier remembers their first firefight. I remember mine — standing behind a griddle flipping steaks while mortars screamed overhead, dust in my teeth, and a tray of MRE corned beef hash threatening to explode in my hands. Welcome to the Flying Spoon Platoon.

The first time I felt real fear was when the company commander called us into the tent for a secret meeting. Whatever confidence I had sank the moment they put the invasion plan for Baghdad up on the projector. It was time to become a man. We were "going to war."

The original plan was a jump-in — fly from Kuwait City, drop over Baghdad International Airport, secure it, and hold it. You mean, we're doing what I actually trained for? What the hell. Fear hit like a cold slap; I was about to get a mustard stain on my parachutist badge. Within 24 hours, plans changed. We scratched the jump and loaded up to push by convoy from Kuwait City through to North Baghdad. Which one was worse?
Didn't matter — time to pull our courage out of the dirt and get gritty.

At that point, we weren't scared of stray bullets — not yet. But the siren they used in chemical warfare exercises could have you scraping stains out of your issued brown underwear by hand. Weren't we there for weapons of mass destruction after all? Whether I heard incoming rounds while holding a spatula or an M4, my career path had officially taken a weird turn.

We crossed into Iraq in late March 2003. The sun rose through dust that made everything look like we were cooking inside a giant hair dryer. Everything was brown — the ground, the sky, the uniforms, even the streaks on our underwear from not showering for weeks. Water was rationed to three bottles a day for drinking, brushing teeth, shaving, and whatever else you needed it for; baby wipes became holy. The only color came from tracer fire and the tiny Tabasco bottles we hoarded like liquid gold.

Our convoy rolled through the border a few days earlier. This wasn't like the training videos. No soundtrack, no clean uniforms, no slow motion. Just miles of nothing, sand crusting and molding your face like a mime frozen mid-expression, and that metallic taste of adrenaline. I remember bodies scorched inside vehicles, plastered to windows like the Crypt Keeper from Tales from the Crypt. Blood stains on walls — head high — with brain matter crusted and dried like spaghetti sauce in a clear Tupperware container. Some images stick in your throat.

We set up our first field kitchen about ten clicks outside Baghdad, near a half-bombed-out training village that looked like time had stopped decades ago. The mission: feed everyone and keep morale up — even when mortars landed close enough to make you jump out of your desert boots, knowing there was nowhere to go.

The Army says, "You're infantry first, then your job."
That's true until breakfast needs to be made for 300
troops and the only rifle you're holding is slung across
your back while you're full battle rattle — flak vest,
helmet, boots — and a long-handled spoon in your hand.

I remember the platoon sergeant yelling, "Peltier! Get
those steaks hot, son! Mortars don't wait for no man!"
"Yes, Sergeant! Flipping under fire!" I yelled back, half
laughing, half-ducking.

We cooked on MKT units — Mobile Kitchen Trailers —
metal beasts whose roofs had to be raised by four men.
They smelled like diesel, grease, and faint terror. Each
had griddles, ovens, and burners that could roast a camel
if you left them on too long. The sizzle of meat mixed
with the distant thump of artillery — that was our
soundtrack.

When supply convoys ran late, creativity was survival.
You invented dishes like "Dirty Dent
Feet" — canned corn, leftover beef, breakfast potatoes,
and a prayer. Or the ever-popular "MRE Surprise":
throw random MRE components into a pot and call it
Jungle Cooking. Pretty comparable to the culinary
creations in county jail, honestly.

One morning we were playing spades, stripped down to
boxer shorts, trying to catch a breeze while babysitting a
weapons cache — 150 feet long and six feet high,
stacked with artillery and munitions. An announcement

went out for a controlled detonation — which we didn't hear. The blast shook the building; my comrade, a Jamaican soldier, jumped up, left everything — including his weapon — and ran out screaming in his underwear. I went from laughing tears to crying as I planned breakfast.

Later, while stirring a vat of reconstituted pancake mix, a mortar landed maybe 200 meters away. The blast shook the griddle; half the batter sloshed onto the floor. That same paratrooper screamed, "Incoming!" while I, covered in pancake goo, muttered, "Well, that's breakfast for the scorpions."

That's when I learned the first rule of being a combat cook: fear comes second, food comes first. Well, for most of us.

We flipped steaks, made pancakes, and cracked jokes because that's what kept us sane. You can't control when the next round hits, but you can make damn sure your brothers get a hot meal. That mattered more than anyone admitted. That's where camaraderie came from.

At night, when things quieted in camp — if only for a little — we'd sit on MRE boxes, share random things from our care packages, and talk about home. For some of the troops that was chewing tobacco and cigarettes. The lengths smokers would go so the red cherry at the end of a cigarette didn't give them away were pure MacGyver — sliding the bottom off an empty magazine and hiding the butt inside. The topics never changed: girls, cars, what we'd eat if we ever saw a real kitchen again. Someone would always shout "steak," and the whole platoon would groan. I dreamed about my dad's lasagna from Sorrento's, thick with layers and real mozzarella. Out there, even powdered mashed potatoes could taste like heaven if you had the right mindset. You had to be creative in thought to get to the next meal.

One of the dumbest things we did involved a K-Kliff, a large metal vat with a modern burner unit used to boil canned food. You're supposed to fill it halfway with water. One sergeant put in barely any water and fell asleep. Hours later, the cans inside swelled and exploded like frag grenades. Beans everywhere. We laughed until the platoon sergeant threatened to smoke the entire "Flying Spoon" platoon. He couldn't hide the grin though.

By mid-deployment I realized food was more than sustenance. It was the one thing that reminded us of who we were before uniforms. Guys would come through the line, tired, dirty, maybe missing a buddy — and when they tasted something warm, you could see normalcy flicker back for a split second.

We were soldiers, sure — but in that kitchen, we were healers, wheelers, and dealers. We just used butter instead of bandages.

ARMY RECIPE — BEEF STEW, FIELD STYLE
Source: TM 10-412 — Modified for
War Zones and Questionable Sanity
Yield: 100 portions (or 85 if everyone
comes back for seconds) Ingredients:

- 25 lbs beef chunks (canned or fresh if you're lucky and
  a truck showed up)
- 5 lbs dehydrated onions
- 10 lbs potatoes, diced
- 5 lbs carrots, sliced
- 2 lbs flour
- 3 gallons bottled water (that has been sitting in 120°F
  heat)
- Salt, pepper, probable sand, and hope to taste
- Optional: 1 smuggled bouillon cube from a care
  package for "flavor enhancement"
  Directions:

1. Brown beef in oil until "almost" cooked. (If it's too
   brown, call it Cajun-style.)

2. Add onions, potatoes, carrots, and stock. Simmer until
   tender or until the next attack.
3. Stir in flour to thicken. Curse whoever used the whisk
   for scrambled eggs and didn't wash it.
4. Serve with humor and a side of situational awareness.

Sergeant P's Note: If the stew sticks to your spoon, it's
perfect. If it sticks to the wall, you're promoted — which
means you get to do it again tomorrow.

Chapter Three: Shit Burning and Steak Turning

I remember the smell before I remember the faces. Diesel, sweat, smoke, and burning shit — all mixed together into a scent you could taste. Aqua Shit-io. It clung to your clothes, your gear, and your skin. You could shower ten times and still smell it. Hell, I think I can still smell it now if I close my eyes long enough. That was the perfume of The Flintstone Cave — our little slice of prehistoric hell on Earth.

The cave was nothing glamorous. It was an old, half blown-out structure that looked like the set of The Flintstones if they'd filmed it in Satan's armpit. Sand colored walls, jagged rock ceilings, and the kind of heat that could melt the chrome off a Humvee. We'd set up camp there like we were starring in Mad Max: The Chow Line Edition.

We would be stripped down to our boxer shorts at night with cots lined up fighting over one fan stuck on rotation. By the time the fan got back around to you, your nuts were stuck to both of your thighs again. You'd wake up drenched in sweat, your underwear glued to your ass, the sound of generators and flics buzzing louder than thoughts. The only relief was the occasional breeze of the fan that blew through — and it always carried the smell of the burn barrels.

The Shit Burn Detail from Hell

Every base had its rotation for shit-burning duty. It was the lowest of the low. You hadn't truly lived Army life until you've stirred a fifty-gallon drum half-full of human waste with a stick, while diesel flames roared up around you like a portal to hell.

"Peltier! Grab the stick, your turn!" yelled the company First Sergeant, a thick-voiced man who sounded like he'd smoked gravel for breakfast, and chewed on rocks for fun.

"Roger that, First Sergeant," I said, holding my breath and trying not to gag as I took my spot by the barrel, holding that stick like I was the last man standing.

The flames danced, the black smoke rose, and the stench — my God, the stench — was enough to make your eyes water and your ancestors curse you.

"Jesus Christ," I muttered, "I'd rather take mortar fire than smell this."

From across the pit, my Jamaican battle buddy paratrooper laughed, wiping sweat from his forehead. "Ya see, mon, that's da smell of freedom! If freedom had a butthole, it'd be dis!" The whole crew cracked up. That's how we survived it — laughing. You either laughed or you puked, and puking just meant double duty.

We'd joke about it like it was a gourmet job. "Hey, Peltier," the sergeant yelled once, "you think Gordon Ramsay could handle this shit?" I stirred the drum, face half-covered with my week-old sweat-stained brown undershirt that had white salt stains all around the neck. "Hell no, Sergeant. He'd call it 'Malt O Meal with hints of regret!'" The boys roared. Gallows humor was the only thing that kept the darkness out. Some say we were sick in the head, and I kind of agree.

The Kitchen Hustle

Once shit burn duty wrapped up, it was time for the next round of hell — cooking. We didn't think it was ideal for the cooks to have to do shit-burning detail, but again, we were shitting in the barrels of hell also.

We'd rigged up our MKT units' right outside The Flintstone Cave. It was already old, probably from the 80s, and smelled permanently of grease, gunpowder, and nightmares. Each one was hotter inside than the desert itself. We cooked in full battle rattle — flak vests, helmets, rifles slung — because incoming rounds didn't give a damn about meal schedules and sure as hell didn't have names on them.

"Peltier! Those steaks ain't gonna cook themselves!" the sergeant barked.

"Yes, Sergeant! Flipping under fire!" I yelled back, spatula in one hand, rifle hanging off my shoulder.

The grease popped, sweat poured, for a little extra salt flavoring, and every so often, the cave would rumble from a blast off in the distance. We didn't even flinch anymore. It was second nature for there to be incoming at this point.

The Jamaican would wander over, wiping his face with a rag. "Ya think da Taliban like steak, mon?"

"Probably," I said, "but they ain't gettin' mine."

He laughed, slapping my back. "You right, Peltier. Ya cook like ya angry!"

"Yeah, well," I said, "the Army taught me how to weaponize frustration."

Sometimes, the supply trucks didn't make it through, and we'd have to get creative. We called it combat cuisine or jungle cooking. You could hand us a can of beans, a piece of jerky, and some instant coffee, and we'd make it taste like Sunday dinner. My personal favorite was Desert Chili Deluxe — MRE chili mac, some Tabasco, a scoop of mashed potatoes, and a dream.

"Peltier, what the hell is this?" the sergeant asked once, staring at a pot of whatever I'd concocted.

"Combat culinary innovation, Sergeant. Michelin-star MREs."

He smirked. "Boy, you're outta your goddamn mind. Keep it up."

The Heat and the Madness

You don't really know heat until you've been in Iraq in full gear, standing over a griddle at noon. The air shimmered. Sweat rolled down your spine in rivers, and you could feel it all the way down to your socks and into your boots.

I saw one guy try to fry an egg on his helmet — and it worked.

At one point, I swear I hallucinated. I thought I saw a Taco Bell sign in the distance. There was no way, a Taco Bell in Iraq. No, it was just a burning Humvee.

We joked our way through the pain. Every second felt like forever, and yet somehow, the days blurred together. The rhythm of war wasn't gunfire or orders — it was the clang of cooking pans, the roar of generators when they didn't run out of fuel, and someone yelling "chow's up!" through a sandstorm.

The Brotherhood of Bullshit

Even in that madness, we had each other. At night, when the heat finally backed off a little — like all the way down to 100 degrees Fahrenheit — we'd sit outside the cave, playing spades and talking shit. The Jamaican would hum Bob Marley under his breath, the sergeant would be cleaning his rifle, and I'd be nursing a half warm Rip-It like it was a fine bottle of rum.

"You ever think about home, Peltier?" the Platoon Sergeant asked once. He was a Sergeant First Class and had been to the first Iraq War in the 90s.

"Every damn day," I said. "But then I think about that smell and realize I'll never forget this either."

He grinned. "Good. Means you're living it right. This is the part nobody tells you about — the stink, the heat, the boredom, and the laughter. You'll miss it someday."

I didn't believe him then. But he was right. You don't miss the war. You miss the idiots you fought it with.

Reflection

The Flintstone Cave was hell, but it was our hell. Between the burn barrels, the grease fires, catching one another on fire using a lantern to fill generators in the night, and the laughter, we built something out of nothing — a rhythm of survival, a weird kind of pride.

Every time I handed a plate to a soldier — dust on his face, eyes heavy from no sleep — I knew that for at least a minute, he'd taste normal again. That was my job. Not just feeding bodies — feeding souls. I was a Combat Chef!

We weren't heroes. We were just cooks, scraping burnt food off pans in a warzone that smelled like death and diesel. But we kept them fed. And that meant something. By the way, don't ever call us a spoon that is for us to call one another.

Army Recipe: Sand-Crusted Ribeye (Desert Steak Deluxe)

Source: TM 10-412 — modified by Sergeant P's Unholy Imagination

Yield: 50 portions (or 45 if the cooks have to taste test several to make sure they're right)

Ingredients:
- 25 lbs questionable ribeye (half thawed, half mystery)
- 2 gallons of pure sweat (literally)
- 1 handful of desert sand (optional but unavoidable) - Salt, pepper, and motor oil aroma to taste
- One dented MKT grill that's seen more combat than most privates

Directions:
1. Scrape the grill with a busted spatula or your Gerber knife until it stops smoking (mostly).
2. Slap steaks down hard enough to scare flies away for 3 seconds.
3. Let sear for 5 minutes or until tracer rounds pass overhead.
4. Flip with precision and a few "fuck this grill is hot" phrases.
5. If sand blows in, call it seasoning.
6. Serve with sarcasm, Rip-Its, and a side of laughter.

Sergeant P's Note: If it ain't bleeding, it's overcooked — and you were getting it overcooked anyway. Suck it up and move on, soldier.

Chapter Four: Midnight Chow and Incoming Mortars

The first thing I noticed about the new outpost ten clicks north of Baghdad International wasn't the dust or the noise — it was the feeling that the air itself was waiting to explode. The whole place looked like a half-finished Lego fort built out of concrete T-walls, sandbags, and desperation. The perimeter wire hummed in the hot wind, and even the shadows looked tired, including my own by this point.

We'd barely been there a week when they told us we'd start running midnight chow with some civilian contractors that were brave enough to already be in country during the middle of a war. This kind of late night meal was meant to keep morale high and stomachs quiet before patrols. It sounded simple on paper. It never was, because I was the one up at night making sure people had something to eat at midnight, and then swinging back around to be up again at 0400 to start breakfast.

The MKT sat near the back of the compound, wedged between a fuel bladder and a line of Humvees that looked like they'd been spit out of hell. The smell was equal parts diesel fumes, burnt oil, and whatever parts and pieces got run over on the road the day before. We had now moved into a giant circus tent. The air shimmered with heat off the vinyl sides of the tent even though it was after 2300 hours.

"Peltier!" my Platoon Sergeant barked. "If I don't get some chow in me before this shift, I'm liable to look like Michael Jackson with a blowback perm."

"Roger, Sergeant," I said, already sweating before I even fired up the burners. "But you're probably going to look like The Rock!" Referring to his uncanny resemblance to the rapper Jay-Z.

My Jamaican brother leaned against the tent post, grinning. "Eh, mon, ya cookin' or ya meltin'? Whole place smell like ass and ambition."

"Both," I said, cracking three full cases of fresh eggs into a giant silver bowl. "If I pass out, flip me before I burn."

We had about thirty guys lined up, night-vision goggles glowing faintly like green insect eyes. Some were laughing, some dead quiet. The tension was there — you could feel it in the way people chewed. Everyone knew the insurgents liked to hit us when the lights were on and our guard was down.

The grease was popping, the radio humming low with static, loudspeakers from the city blaring what sounded like Iraqi karaoke night, when — POP! Right outside of the DFAC, there was a shot. False alarm — it wasn't the enemy, it was another private clearing his weapon into the barrel without dropping his magazine. A round exploded into the sand-filled bottom of a fifty-gallon drum.

Shortly after this incident, there was a large BOOM. The shockwave punched through the tent like a giant rubber fist. The lights flickered, dust fell from the ceiling, and everyone froze for a heartbeat.

Then chaos.

"INCOMING!" someone yelled.

Another whump — closer this time. Dirt sprayed across the kitchen like confetti from hell. The Jamaican hit the deck, covering his head. I ducked behind the griddle for half a second — instinct — then went right back to cracking eggs.

"Peltier, what the fuck are you doing!?" the Sergeant yelled, diving behind a pallet of water.

"Breakfast, Sergeant!" I shouted. "Mortars don't cancel chow!"

The Jamaican peeked up from the floor, eyes wide. "You mad, mon! The sky droppin' bombs and you crackin' eggs?"

"Can't let them catch me slippin'," I said, half-laughing, half-praying. Okay, I was praying.

The third round landed far enough away to just shake the tent. We'd gotten lucky — either they were blind, drunk, or just celebrating. The whole camp was yelling orders, boots pounding, and radios crackling. And in the middle of it, I was still cooking, the wire whisk clanging against the steel.

I'll never forget that night. It was the scent of insanity, courage, and breakfast.

A few of the soldiers crawled back toward the serving line, laughing like maniacs. "Hey, Peltier," one said, "if I'm dying tonight, I want one last omelet — and I want my eggs fertilized!"

"Roger that," I said, tossing it on the griddle. "How you want it — scrambled or scared like you?"

Another soldier popped up, face covered in dust. "Goddamn it, Peltier, you're outta your mind!"

"You're not telling me anything I don't already know, Sergeant," I said. "Now grab a plate before its cold."

Even the Jamaican started laughing again. "Ya man, if we go, we go full-bellied. I want to be so full I can't even see my dick!"

When the all-clear finally sounded, my hands were shaking — not from fear, but from adrenaline. My heart was racing like a generator on overdrive. The reality of what just happened never really hit until later.

We stepped outside the tent — the smell of burnt metal and sand filled the night. In the distance, a few fires glowed like candles on the horizon. The Sergeant looked over at me, then at the griddle still sizzling behind us.

"You kept cooking through all that," he said, half a smile breaking through the grime.

"Yeah," I said. "Wasn't about the food. It was about keeping us moving. If we stop, we start thinking — and when you start thinking, that's when bad things happen."

He nodded. "You're a sick bastard, Peltier. But I'll be

damned if that wasn't the best omelet I ever had."

The Jamaican chuckled. "Ya see, mon? Food make

da soul bulletproof." And he wasn't wrong.

That night, we served over a hundred eggs and omelets, a few burnt steaks, and enough instant coffee to keep a platoon awake through Armageddon — Baghdad edition. Every bite was a reminder that even when the sky was falling, we were still alive, still laughing, and still human.

Reflection

That night taught me something I didn't learn in Basic Training or Airborne School — that courage isn't always about charging toward the enemy. Sometimes it's about flipping food and the middle finger up while the world explodes around you.

You learn to find peace in the noise. The rhythm of cooking, the hiss of the griddle, the laughter in the face of fear — it became my heartbeat. It was how I fought back against the chaos.

People think war is just about bullets and bravery. But sometimes, it's about omelets and burnt steaks at midnight — the small things that keep men grounded when everything else goes to hell.

We weren't heroes. We were just hungry bastards refusing to quit.

Army Recipe: Combat Omelet — Incoming Mortar
Edition
Source: TM 10-412
(Modified for Explosive
Conditions) Yield: 100
omelets (or 75 if a quarter
of the batch hits the dirt)

Ingredients:
- 10 lbs dehydrated egg mix
- 2 gallons bottled water (lukewarm, sand optional)
- 1 can of Spam, diced (if it hasn't been raided)
- 1 bag of powdered cheese (aka orange regret) - Salt,
  pepper, and adrenaline to taste

Directions:
1. Mix eggs and water until smooth — or until the next
   mortar lands.
2. Pour onto a hot griddle, preferably one that's not on
   fire.
3. Add Spam and cheese while praying to whichever god
   handles cooks.
4. Flip once, swear loudly, and serve fast.

Sergeant P's Note: If it's burnt, it's "Cajun." If it's raw,
it's "tactical." If you're still alive after eating it, it's a
success.

Chapter Five: Convoy Cooks

The morning of the convoy started like any other — hot, gritty, and heavy with that quiet tension that sat on your shoulders like a loaded ruck. The air outside Baghdad smelled like a mix of burning plastic from the burn pits a hundred meters behind our sleeping quarters, diesel, and the ghost of yesterday's chow seeping out of my ass cheeks. I had just finished loading the last of the food crates into the Humvee when my Platoon Sergeant walked up with his usual morning bullshit. "What's up, Pelt?"

"Peltier," he said, squinting at me, "you know why you're on this run?"

"Because you hate me, Sergeant."

He smirked. "No, because no one else is dumb enough to volunteer."

"Guess that makes me a patriot and an idiot," I said, slamming the tailgate shut.

The Jamaican paratrooper was already inside the passenger seat, humming something under his breath and loading his rifle. "Ya see, mon," he said, grinning, "today we ride with destiny. Or maybe we ride with diarrhea. Either way, we goin' north, and maybe something goes south."

That was the thing about convoy runs — you never knew what waited for you outside the wire. Could be clear roads, could be chaos. Either way, once you left that gate, you weren't in control anymore.

Leaving the Wire

We were to roll out before sunrise to deliver chow. I recall six Humvees, two five-ton trucks, two-and-a half-ton trucks, all soft skin and no armor — one mission: deliver food to an outpost farther north. The sound of engines drowned out the morning prayers echoing from the city mosques. I always found that eerie — one side praying, the other side loading magazines.

The roads were shitty and full of half-buried trash. Every pothole looked like a bomb, every abandoned car like a trap. You'd stare at a pile of dirt for miles, waiting for it to explode, while trying to stay in the middle and away from the sides.

The Jamaican broke the silence. "Eh, Peltier, ya think them camels get hazard pay?"

"Only if they hump the person riding them," I said, keeping my eyes forward.

The Sergeant chuckled from the lead vehicle over comms. "Focus up, soldiers. If you see a goat holding a cell phone, shoot it."

Everyone laughed. That was how we dealt with fear —
we turned it into punchlines. If you couldn't laugh at the
madness, it'd eat you alive.

The Calm before the Punchline

The sun was climbing, turning the sky into a white
furnace. I leaned against the Humvee, sipping a warm
bottle of water, dust grinding between my teeth like
sandpaper. The heat started to rip down into the back
seat from the gunner standing through the roof of the
Humvee on the 50-cal gun.

The Jamaican said, "Ya know, mon, I miss da smell of
Da Ocean. Not this."

I nodded. "Yeah. Smells like death, diesel, and asshole."

He laughed so hard he almost dropped his rifle and
started talking about the ocean and his ex.

The Platoon Sergeant in the front seat shook his head. He
could turn serious situations off and on at the snap of a
finger. "Stay alert. Intel says there's been some
movement near the next checkpoint. Don't let your guard
down and keep your head on a swivel."

As we continued rolling, eyes scanned every rooftop and
trash pile. That's when I felt it — that shift in the air.
Soldiers get this sixth sense. You just know when
something's wrong and not feeling right.

Impact

It happened fast — too fast to think, right in front of us.

The dump truck came out of nowhere, barreling from a side road. It wasn't speeding toward us — it was swerving like the driver was drunk or panicking. The Sergeant yelled into the
comms, "HOLD YOUR LANE! HOLD YOUR LANE!"

But the truck didn't stop.

It clipped the front Humvee — not a full hit, more like a shove — and then tipped. The weight of dirt and rock collapsed the thin roof like a soda can with my fellow soldiers inside.

The sound — Jesus, the sound — metal crumpling, men yelling, the ground shaking. We slammed brakes, sand and dust filling the air like smoke.

The Jamaican was out of the Humvee before I could stop him, sprinting toward the wreck. I followed.

The dump truck was on its side, steam hissing, and one tire still spinning. The Humvee was half buried. The dirt had poured over it like a grave. We dug with our hands, calling out, coughing from the dust and fumes.

The Platoon Sergeant was barking orders, voice cracking. "Get a perimeter up! CALL IT IN!" But deep down, I already knew. You could feel it in the silence between the shouting — that awful stillness that follows chaos.

We pulled one man free — broken but breathing. The other two... no. One of these soldiers had been teaching me to play drums on his drum pad just hours before the convoy and had let me use his Sony Walkman to listen to the 50 Cent album that had dropped just before our deployment.

We stood there, covered in dirt and sweat and someone else's blood, staring at the wreck. The Jamaican dropped to his knees, muttering a prayer in patois. The Sergeant took off his helmet, eyes hard.

"FUCK," he said quietly.

That's all it ever was out here. Luck and timing.

The Ride Back

The convoy continued to the base to deliver chow, slower than before. Nobody spoke. The radio was quiet except for static. The sun started to dip, painting everything in gold, like the desert was mocking us — beautiful and cruel at the same time.

Chow delivered and convoying back to our home base felt eerie, like time had stopped.

When we got back, the kitchen was waiting — like nothing had happened. Same grease stains, same smell of old food and onions. I stood there staring at the griddle, still crusted from the morning breakfast, and felt like the world had split in half.

The Jamaican finally broke the silence. "Ya know, mon," he said softly, "we deliver food, not bodies."

"Yeah," I said. "But today, the menu changed, and we have to annotate that on the recipe card and adjust fire."

He looked at me, nodded once, and went back to work. Because that's what soldiers do — we keep moving.

Reflection

That night, I sat outside the tent with a near-beer I had swindled a civilian contractor into giving me, and I sat in silence. My Platoon Sergeant joined me, not saying a word for a long time — he was always one to speak and never at a loss for words.

Finally, he said, "You did good out there, Peltier. I'm proud of you."

"Didn't feel like it."

He nodded. "It never does, Pelt; it never does."

The stars above Baghdad were clear that night. You could see forever. Funny thing about war
— It makes the sky look bigger. Makes everything else feel smaller. I remember looking at the Big Dipper and thinking back to my childhood when I would look out at the stars and wonder what else was out there. This — this is what was out there: war and misery.

Luck was our religion. Every day you survived was a prayer answered. But that day, two good men lost their faith.

Army Recipe: Convoy Chili — Road Dust Edition
Source: TM 10-412 (Modified for IEDs and Bad Roads)

Yield: Enough for 100 hungry bastards and several broken hearts.

Ingredients:
- 20 lbs ground beef (or whatever hasn't spoiled; you might have to stretch it a bit)
- 10 cans of beans (assorted mystery beans welcome)
- 5 lbs diced onions (dehydrated, because the fresh ones were dehydrated already anyway)
- 3 lbs tomato paste
- 2 gallons of bottled water, preferably not boiling hot from the sun
- One handful of sand (optional, but inevitable) - Salt, pepper, Tabasco, and survivor's guilt to taste

Directions:
1. Brown the beef in a pot balanced on a generator exhaust pipe, or on a stove if you actually have one.
2. Add beans, onions, and tomato paste while keeping one eye on the horizon.
3. Stir with a bayonet from your M4 if you've lost the ladle.
4. If dust blows in, that's just "smoked chili powder."
5. Serve hot, because everyone loves hot food when it's already hotter than Satan's ball sack outside.

Sergeant P's Note:

If you are crop-dusting the Humvee from the turret, it's chili. If it runs down your boots, it's art. Either way, eat fast — we roll at dawn.

Chapter Six: Green Zone Mirage

Mid-war. Baghdad Airport secured, Saddam's Palace secured — what is this? Rest and Relaxation? No, we weren't going home, but we got a day trip to the Green Zone. I swear, stepping into the Baghdad Green Zone for the first time felt like being teleported to another damn planet. One minute you're driving through craters, burnt-out cars, and half-dead palm trees — the next, you're driving under the two large swords across the road and then standing in front of Saddam Hussein's Republican Palace, and it's shining like the world didn't end.

Marble columns, gold trim, a goddamn fountain running, and a full pool in the back like there wasn't a war two miles away. I even have a photo on the front page of the Muskegon Chronicle on Saddam Hussein's gold couch.

The smell hit me first — not the usual cocktail of ball sweat and burnt shit. No. This smelled like chlorine, soap, and spoiled privilege. The kind of scent that said, "Relax, soldier. We don't bleed here."

We were there to pick up supply manifests and eat a hot meal. Half a day of luxury. That's all.
Just long enough to remind us we didn't belong and to get the hell out.

The Platoon Sergeant smirked as we stepped off the truck. "Welcome to paradise, boys. Don't touch anything, especially the females. Don't stare too long, and if someone offers you shrimp — take it, even if you're allergic."

My Jamaican brother was already wide-eyed, spinning in a circle. "Ya see dis, mon? Look like a damn movie! You sure we ain't dead?"

"Not yet," I said. "But I think this is what purgatory smells like."

The Palace of Kings

Inside was like walking through a fever dream. Marble floors so shiny you could see your filthy face staring back at your freshly starched combat uniform. Of course, they weren't really starched and pressed, but they were stiff from sweat and stank, and if you took your uniform off, it would stand up on its own like Casper jumped inside. Chandeliers hung like they were stolen straight from Vegas. There were murals of Saddam himself — the guy really loved his own face. Most other portraits around town were destroyed by locals and soldiers, but this was different.

We walked past soldiers and contractors lounging on comfortable, cool leather couches, air conditioning blasting cold air that felt like sin. I was used to the hot wind ripping across my face, looking like the Joker from Batman when I stopped.

Then we hit the pool.

I froze. My brain needed a reboot. For the first time in months, I saw women — actual, live women — in bikinis. Sunbathing, laughing, and sipping on drinks like they were at a club in Miami. Some were civilian contractors, a few were female soldiers taking advantage of the break. Every single enlisted guy walking by turned into a Neanderthal.

The Jamaican nudged me, whispering, "Mon... dis can't be real. Is this Heaven?"

"Nah," I said, "If it was Heaven, we'd be the ones in the pool."

He grinned. "Ya right. These officers out here look like kings, mon. Kings of bullshit."

He wasn't wrong. Officers had taken over the place like it was their own damn country. Some were shirtless, wearing Oakley's, and decked out in all the top gear they'd swindled from their supply company. Drinking imported beer, lounging on Saddam's deck chairs. One lieutenant colonel strutted by in swim trunks with a cigar in his mouth like Tony fucking Montana.
Meanwhile, we still had sand in our ass cracks and grill smoke in our lungs.

"Must be nice," I muttered. "A palace for the privileged."

"Ya think they got a bathroom without flies?" the Jamaican joked. "Maybe I shit like royalty today and not have shit-burning detail."

The Poolside Photo Ops

We weren't supposed to be taking photos, but every one of us had those little wind-up disposable cameras from PX or packages that made it to us months after being sent by family. You could hear the faint click-whirrr of shutters all around as soldiers pretended to pose — thumbs up, big smiles — while secretly angling their cameras just right to catch a bikini in the background for the spank bank later that night.

"Hey Pelt," the Sergeant said, squinting at me, "don't get caught creepin'. The last thing I need is you explaining to First Sergeant why you're the Ansel Adams of Ass."

I laughed. "I'm just documenting morale, Sergeant. Strictly historical, enjoying the moment."

We posed in front of the palace pool, pretending to admire the fountains while snapping blurry shots of tan lines and water glistening off curves. Juvenile? Hell yes. But after six months of sand, sweat, and dudes... it felt like witnessing a miracle. We hadn't seen a single female up until this point, coming from an all-male infantry unit with the 82nd Airborne Division.

The Chow Hall Feast

Then we hit the Green Zone chow hall, and that's when it got really surreal. This wasn't powdered eggs and grilled sizzler steaks. We're talking T-bone steaks cooked to order, shrimp cocktail, mashed potatoes with real butter, and cold Coca-Cola.

I swear, I almost cried when I saw ice cream bars in a freezer. We piled our trays like savages. I was gluttonous and ate like I hadn't seen real food in a year — because I hadn't. I was 220 pounds pre-deployment while still fit; I had dwindled down to about 178 pounds. Every bite felt like cheating.

I caught myself feeling guilty halfway through a plate of fried chicken and scalloped potatoes, thinking of the boys back at the outpost choking down MREs.

The Sergeant leaned over and said, "Don't think about it, Pelt. Just eat. They'd want you to."

"Yeah," I said, "but I bet they'd want some of this too."

He laughed. "Don't worry. You'll shit it out before we hit the gate."

Souvenirs and Swaps

After chow, we wandered through the Green Zone marketplace — stalls run by locals selling bootleg DVDs, fake Rolexes, Saddam's stolen ashtrays, scorpions encased in glass, and whatever else they could scavenge. Soldiers traded Rip-Its, Copenhagen, cigarettes, and MRE peanut butter — or occasionally faintly swayed their M4 toward a vendor for a "free gift."

The Jamaican somehow bartered two cans of tuna and a half-pack of crackers for a Saddam-branded money clip. "Look at dis, mon! The dictator's fishing for more money!"

I scored a "Sir, very good DVD quality" of Friday After Next with Chinese subtitles for a Rip-It and a handshake. Felt like a fair trade until people started walking across the screen when we finally watched it.

The Green Zone wasn't just a place — it was a mirage. The illusion of safety, comfort, and luxury, floating above the real war like an air-conditioned dream.

Back to the Dirt

When it was time to leave, none of us said a word. Stomachs full, minds heavy. The Humvee engines started up, and I watched the palace shrink in the side mirrors. For a minute, I thought about what it'd be like to stay there — sleep in a bed, drink cold water, maybe even forget we were in hell.

But the thought didn't stick.

By the time we hit the first checkpoint outside the wire, the air changed again — hot, heavy, familiar. The silence returned, and I felt my shoulders drop like I could breathe again.

"Ya miss da pool already, mon?" the Jamaican asked.

I shook my head. "Nah. That place wasn't real. Shit was a fantasy."

The Platoon Sergeant grunted from the front seat. "Real's overrated, Pelt. But I get it. Comfort makes you soft. Dirt keeps you sharp."

He wasn't wrong. The Green Zone had been a fever dream — bright, beautiful, and fake as hell. Out here, where the air burned, at least it was honest.

Reflection

That night, back at the base, I sat on a box of MREs tilted upright, staring up at the desert sky as I usually did. No lights, no fountains, no bikinis — just stars and silence.

The palace already felt like it happened to someone else. Maybe it did. The version of me who laughed at the pool wasn't the same one who'd crawl into a cot tonight listening for mortars whistling through the air. Hell, I was probably putting on headphones and listening to

Alicia Keys to ease my mind anyway, so I didn't hear a thing.

The Green Zone was a mirage — and I realized I didn't need paradise. I just needed purpose. We didn't belong with the kings. We belonged in the chaos. We were the Flying Spoon Platoon, and we were fueled by fire.

Army Recipe: Luxury Lasagna — Green Zone Edition
Source: TM 10-412 (Modified for Palatial Bullshit
Conditions)
Yield: Enough for 40 officers, 10 contractors, and 10
dusty cooks with bad attitudes.

Ingredients:
- 10 lbs ground beef (fresh, not freeze-dried — you lucky
  bastard, our truck came in today)
- 5 lbs real mozzarella cheese (the kind that actually
  melts and doesn't have plastic particles)
- 4 lbs ricotta (luxury item, handle with reverence)
- 3 jars marinara sauce that doesn't taste like ass
- 2 lbs lasagna noodles (boiled in clean water — a
  miracle)
- Salt, pepper, and delusion to taste

Directions:
1. Layer beef, noodles, and cheese in a real oven — yes,
   a real oven, not a steel box heated by a blowtorch.
2. Bake until golden brown or until a contractor wanders
   off with your tray.
3. Serve with cold Coca-Cola and a side of survivor's
   guilt.
4. Eat it fast before you're back to powdered eggs and
   MRE crackers on a mission.

Sergeant P's Note:

If your lasagna doesn't make you feel like Garfield and question your life choices, you're not doing it right. Enjoy it one layer at a time — tomorrow, it's back to the dirt.

Chapter Seven: Mail Call and Memories

If you ever want to see grown men act like kids on Christmas morning, watch a mail call in Iraq. The minute someone yelled, "MAIL!" over the radio, every soldier within earshot appeared out of nowhere — half-dressed, dirty, limping, running, tripping, like cockroaches in a filthy kitchen when the lights flicked on — because getting a box or letter from home was better than your next meal.

The stale air from back home was trapped inside those packages — a mix of excitement and desperation. Guys who hadn't smiled in weeks suddenly looked alive again. I remember the way it felt holding a box with my name scribbled in my mom's handwriting — that shaky, looped script that looked like comfort itself. I never understood how someone could have such perfect cursive handwriting.

My mom had me locked in; after all, I was her baby boy. Every Friday, she'd send me a copy of The Muskegon Chronicle, no matter what. Front to back, folded neatly, sometimes with a handwritten note in blue ink: "Stay strong, Nate. Love you. GO ROCKETS!" (Reeths-Puffer High School, baby — football was religion back home in West Michigan.) She would also send subtle reminders that if she hadn't heard back from me, she would perform an out-of-body experience and come see me.

She was into that Sylvia Browne, psychic, outer-body experience stuff. I said, "Mom, stay your ass right where you are. I'll be fine," because if by any chance it was real, I didn't want any part of it.

By the time I got the papers, they were always three weeks late. I'd be reading about touchdowns and rivalries that had already been settled, like watching a rerun of my own hometown — and, of course, the Muskegon Big Reds were dominating. Still, I read every damn word — scores, ads, obituaries, the police blotter. It made me feel human again. Like I was still tethered to the world I came from, even if the cord stretched across the desert like an umbilical cord.

One of the guys asked once, "What's black and white and red all over, Peltier?"
"Because," I said, flipping the page, "I don't know — a dying zebra?"
He nodded. "Fair enough, but a newspaper, you idiot — you're in the middle of the desert reading your hometown paper thousands of miles away."

Sometimes, the newspaper came with other treasures — a bag of homemade cookies turned into crumbs, beef jerky, sunflower seeds, or letters that smelled like her old Avon perfume. Other times, there'd be random clippings — obits, funny comics, or even grocery coupons.

She'd circle things and write dumb notes like, "Thought you'd laugh — ham is on sale at Meijer!" I did laugh.

Then I'd fold it up and tuck it into my Bible or stuff it under the netting of my Kevlar helmet. Why in the hell would I care about ham on sale at the local grocery store?

But the weirdest mail Mom would send — and my favorite in a sick and twisted way — came in the form of Busted Magazine.

That little magazine was infamous in Muskegon County. It was basically a weekly mugshot parade of everyone arrested or convicted. You'd flip through it like a yearbook from hell. And every time I got one, I'd recognize at least three faces.

There was the guy I used to play baseball with who got caught stealing paperclips from Meijer, the neighbor's kid who swore he'd never get caught stealing stereos out of cars — pretty sure he got mine. Some were DUIs, some assaults, some just looked like they'd run out of luck.

The first time I saw it, I sat there in the tent laughing my ass off. "Look at this shit," I said, showing the Jamaican and my Platoon Sergeant. "This dude used to sell CDs out of his trunk at Ray's Mini Mart. Now he's in the county for retail fraud and breaking and entering."

The Jamaican took one look and started laughing so hard he fell backward on his cot. "Ya see, mon? Everyone fightin' a different war."

The Platoon Sergeant smirked. "Hell, we're all brothers — they get three hots and a cot. We get three hots and a cot."

We'd pass the Busted Magazine around the tent, calling out names and mocking them like it was a comedy night. Some of the guys would recognize people too — "Damn, that's my cousin!" — And we'd lose it.

But once the laughter died down, the quiet would hit. Because it was impossible not to think about how messed up it all was. They were stuck in jail cells back home, and we were stuck in a sandbox halfway across the world. Two different prisons, same damn cage. One just had sand and mortars instead of concrete and bars.

I remember sitting outside the airport hangar one night, reading the Chronicle by a tiny red-lens flashlight while the generators hummed. I flipped a page and saw a photo from back home — the Muskegon Big Reds under those Friday night lights, helmets shining, the crowd roaring. You could almost hear it if you closed your eyes. I went to a rival high school, but the Big Reds and Hackley Stadium were legendary. The drumline echoing through the tunnel, the Native American mascot riding out on horseback before they banned it — that was home.

And for a second, I wasn't in Iraq anymore. I was back in the bleachers, hoodie on, hot chocolate in hand, freezing my ass off but surrounded by people who didn't have to look over their shoulders every five minutes.

Then a mortar boomed somewhere in the distance, and reality punched me in the gut again. War had a way of doing that — giving you a glimpse of peace just long enough to make the chaos sting worse.

The Jamaican walked up, belt undone, pants halfway off his ass, new boot goofin' as usual. "Whatchu readin', mon?"

"The Chronicle. Big Reds won again."

He grinned. "Good. Means da world still turnin', eh?"

"Yeah," I said. "Just wish I was turnin' with it."

He nodded and handed me a non-alcoholic beer he had stashed away that was piss warm.

"To da Big Reds, mon."

"To home," I said, clinking cans.

We drank in silence, looking out into the darkness ahead with nothing in sight. Sometimes I wondered which was worse — being locked up like the guys in Busted, or being free in a place that didn't feel real. Out here, we were alive, but we weren't living. Every day was borrowed time. There was always that fine line, and we didn't know when the thread would snap.

And yet, when mail call came — when I smelled those old newspapers inside that brown manila folder and saw my mom's handwriting — I remembered who I was before the uniform.

That was enough to get me through another mail call day.

Army Recipe: Comfort Cookies — Mail Call Edition
Source: TM 10-412
(Modified by Desperate
Morale Cooks) Yield: Enough
to make 20 soldiers cry and
laugh at the same time.

Ingredients:
- 4 cups crushed cookies from a care package or MRE
  (bonus points if they arrived as crumbs)
- 1 cup peanut butter from MREs
- ½ cup sugar packets scavenged from coffee rations
- A splash of dehydrated milk for moisture - One
  generous serving of homesickness

Directions:
Mix it all together with whatever you've got — a spoon,
your hands, or pure emotion inside your canteen carrier
that was also used for holding your shaving water. Shape
into balls or sad little pancakes.
"Bake" on the MKT griddle until slightly crispy or until
your Sergeant asks what the hell you're doing. Serve
immediately — share stories of home while pretending
you don't miss it. Sergeant P's Note:
If you eat two quickly, you'll think of home. If you eat
three, you'll wish you never left.

Chapter Eight: Sandstorm Sundays

They say Sunday's the Lord's Day. Out here, in the middle of Iraq, it could be argued as the devils instead — a day that started calm and ended with a mouthful of dust, a sore ass, a weapon you had to clean with a Q-tip for four hours afterward, and a story you couldn't make up if you tried.

That morning, the sky had that strange yellow tint that only happens before all hell breaks loose. It wasn't just heat haze — it was a warning. The horizon shimmered like a mirage, and even the wind felt like it was holding its breath. Something on the edge of the horizon was about to come over us like a bad joke that never landed.

The chaplain's assistant and the chaplain came for the weekly Sunday service, which I always looked forward to. The chaplain's assistant was a brother in his early twenties who was raised in the church choir. His crisp, soulful voice would give you chills as he belted out songs that sounded like they should have been on one of the Columbia House record CDs I ordered months prior without paying for. Yes, I asked for forgiveness — give me a break. I wasn't proud, but I'm certain no one ever paid for one of those CDs to this day. This gave me the motivation to get through what was about to come — or at least I thought it would.

We were out at the range, supposed to be doing weapons-proficiency training, which really meant we were baking under the sun, waiting to shoot at targets that looked like they were drawn by a drunk third grader. The range smelled like simmered sewer rat and armpit stew — the kind of air that sticks to your tongue and makes your breath stink.

The Jamaican was next to me, pretending his rifle was a guitar. "Ya mon, music of freedom, yah?"
"Yeah," I said, "until that Sergeant First Class hears you and turns it into the music of pain." He laughed — right before the shot went off.

It wasn't mine, but nobody cared. Somewhere down the line, some dumbass decided to get cute and fire one off before the command. A single pop — and that was all it took. I knew this was about to be a day from hell while already being in hell.

The Sergeant First Class's voice cut through the air like a buzz saw. "LOCK 'EM UP! EVERY GODDAMN ONE OF YOU!"
You could feel the fear ripple down the line like a wave. Everyone froze. The sergeant stalked toward us, red faced, veins bulging, like a pissed-off pit bull ready to charge.

"WHO THE HELL FIRED?" he barked.

Silence. Not one sound. Even the flies stopped buzzing, which is unheard of. We were stuck in the Matrix.

He looked at me — probably because I was smirking. I couldn't help it. I had that dumb 19-year-old, invincible grin plastered on my face.

"You think this is funny, Peltier?"

"Only if it's not me that fired, Sergeant."

"Wrong answer!"

He pointed toward the dirt. "You just volunteered for a front-row ticket to humility. Duck walk. One hundred meters down and back. Now."

I shrugged. "Roger that, Sergeant," I said, with a shit eating grin on my face.
I dropped down and started the duck walk for about twenty meters as everyone else sat back and watched me. Heavy desert boots dragging through the thick sand, legs already shaking like Hulk Hogan in a WrestleMania match getting ready to rip his shirt off. But I wasn't about to give him the satisfaction.

Instead, I popped up, threw my hands in the air, and started crip-walking — flip around, then moonwalk, killing it too, like a "Smooth Criminal." Then mixing in a few dance moves like I was performing for the USO. The guys were losing it — laughing so hard they could barely breathe. Even the Sergeant First Class tried to hide it behind his hand.

"Peltier!" he yelled. "You think this is a game?"
"Only if we are winning Sergeant!"

He doubled my distance. Two hundred meters down, two hundred back. By the time I was done, I looked like a dehydrated ostrich with PTSD. My knees felt like pudding, and I was sweating from places I didn't know could sweat. I had just enough energy left for victory dances — I shook my legs from side to side as fast as I could, hunched over and flapped my arms like a bird while jabbing my neck back and forth.

"Next time you smart off," he said, "you'll be crawling instead."
"Copy that, Sergeant," I said, still grinning. "At least I'll look good doing it."

The Jamaican just shook his head. "Ya mon, you fuckin' crazy. But I like it."

That afternoon, the real punishment rolled in. The sky went from yellow to red to black — the kind of storm that makes the earth vanish. The wind hit first, then the sand — billions of tiny razors slapping every inch of exposed skin. I tried to stash my M4 in a large plastic garbage bag that I, for some reason, had in my rucksack so I wouldn't have to deep-clean it for an entire day.

You couldn't see five feet ahead. The whole outpost disappeared into a brown blur.

"SECURE YOUR SHIT!" someone yelled. I couldn't even see who it was through all the sand and wind.

We stumbled through the chaos, grabbing our gear as rucksacks flapped like they were trying to take off, and the soldiers were gasping for air sounding like dying hyenas. I was choking on dust, eyes burning, and sweat turning to mud on my face.

The Jamaican was shouting next to me, voice muffled through his shirt. "This no storm, mon! This is Satan's sneeze!"

We finally made it into the kitchen tent, which was somehow still standing. Sand was blowing through every seam. It coated everything — the tables, the pans, even the damn butter. I swear I had grit in my teeth for a week.

"Chow still on?" the Sergeant First Class yelled over the wind.
I looked at the griddle, already covered in dust. "Yeah, Sergeant. Today's special — Desert Grit Grits. Sand included!"

He cracked a rare smile, tongue between his teeth. "Don't scorch them, Michelin Chef."

We cooked through that storm like maniacs. The tent rattled, the roof threatened to lift off, and we were still flipping food like it was just another day. You couldn't tell where the food ended and the desert began. I'm pretty sure I swallowed some bug protein that day along with all the grit.

When the storm finally died down, the world was quiet. Not peaceful — just silent in that eerie post-chaos way. Everything was buried. Trucks, gear, even part of the ammo crate line. It was like the desert had tried to erase us and then changed its mind halfway through. I thought I was finishing filming a scene for The Mummy with Brendan Fraser.

I sat down outside the tent, exhausted, sand crusted on my face. The Jamaican dropped beside me, both of us looking like powdered doughnuts, but I couldn't stop laughing at him looking like Tyrone Biggums from the Chappelle Show.

"Ya ever wonder why we still here, mon?" he asked. I wiped my eyes. "Because we're too dumb to quit."
He laughed softly. "Nah, mon.
Because God need
entertainment." He wasn't wrong.

The Sergeant First Class walked by with his Sony Walkman headphones on, uniform top off with just a brown shirt, pop-lock dancing, looked at me and said,

"Peltier — next time you feel like moonwalking, make sure it's away from the range line and we can battle."
I laughed and replied, "Roger that, Sergeant. But you can't lie — I was gettin' it!"
He grinned. "You're a pain in my ass, but at least you're a funny one."

Reflection

That night, the storm had passed, leaving the air thick and still. The stars peeked through again as they usually did. My throat hurt, eyes burned, and I could still taste dust every time I swallowed.

But that's the thing about Iraq — it had a way of humbling you. Whether it was a Sergeant smoking your ass on the range or Mother Nature reminding you who was boss, you learned to laugh your way through it.

Army Recipe: Desert Grit Grits — Sand Included
Source: TM 10-412 (Modified for Post-Apocalyptic
Conditions)
Yield: Enough for 75 sand-eating maniacs

Ingredients:
- 10 lbs instant grits (can substitute 5 lbs of genuine Iraqi
  sand to equal parts grits)
- 2 gallons bottled water (preferably label ripped off and
  bottle sandy)

- 1 lb butter or oil (if you can find it under the layer of
  dust you had to scrape off it) - Salt, pepper, and despair
  to taste (the only allocated seasonings for the military)

Directions:
1. Mix grits with water until thick enough to stick to your
   ribs, as Mama would say.
2. Stir with your weapon-cleaning rod or bayonet
   (improvise, adapt, and overcome).
3. Add butter and sand (don't fight it — it's there
   anyway).
4. Serve hot, crunchy, and slightly bitter — just like the

   war. Sergeant P's Note:

If it crunches, it's seasoned. If it scratches your throat,
it's authentic. If it doesn't kill you — congratulations,
you're officially airborne tough.

Chapter Nine: The Burn Pit Baptism

They say hell has nine circles — I'm pretty sure the tenth one was the burn pit behind our sleeping quarters. About 100 meters back, a black mountain of chaos smoked day and night — never sleeping, never dying. It was our own personal volcano of poison — a reminder that even when the fighting stopped, the air itself was still trying to kill us, and probably still will.

Thus, the reason for the Burn Pit Registry that circulates now for disability claims.

You could smell it before sunrise — that thick, choking cloud that wrapped itself around your throat like an angry ghost. Melted plastic, Styrofoam, scorched rubber, and God knows what else. It burned everything: Humvee tires, old radio batteries, food scraps, and the occasional piece of broken equipment the Army didn't feel like dealing with. "If it fits, it burns." That was the rule.

The First Sergeant used to say, "That smell builds character." He looked like a mix of Mr. Burns from the Simpsons and Colin Mochrie from Whose Line Is It Anyway. I said, "Yeah, well, it's building tumors too." He didn't laugh — but the Jamaican did. "Ya mon, this air thicker than a stripper's ass cheeks with a butterfly tattoo from Sharkey's on Bragg Blvd," he'd say, fanning the smoke with his patrol cap. "Breathe deep, get dat burn pit blessing."

We started calling it the Burn Pit Baptism — our dark joke. Every soldier who survived their first week near that pit was "baptized." You couldn't run from it. You couldn't hide. It crept into your pores, your gear, and your food. We'd dip our fingers into the cap of our canteen and flick water into one another's faces, acting as if it were holy water and blessing each other.

At night, the flames would dance — orange and blue — making me think about the Maize and Blue, the Michigan Wolverines back home. We'd sit nearby, throwing random stuff in, watching things go up in smoke. The black plumes curled into the sky, blotting out the stars. It looked almost beautiful — if you ignored the fact it was pure toxic death, and it was giving our position away to the enemy.

One night, I tossed in an old stack of broken utensils and watched them twist like silver snakes. The Jamaican threw in a few old magazines, one with Vida Guerra on the cover, shook his head, and said, "Goodbye, my sweet Maxim girls. Ya gone but not forgotten." We all cracked up because he always said the most random shit in his broken English.

But every now and then, when the fire got too high, the smoke would roll right toward the sleeping quarters. You'd hear everyone hacking like a tuberculosis ward, trying to cover their faces with shirts. One guy wrapped a pillowcase over his head like it was a gas mask. Another tried to clean the floors with three bottles of bleach and whatever chemicals he found — only made it worse. Probably mixed it with ammonia, because we had to evacuate for two hours before it cleared.

The Platoon Sergeant stormed in, yelling, "You smell that? That's freedom! Breathe it in!" "Bullshit," I coughed. "Freedom tastes like burnt condoms and despair!" He walked off shaking his head, bobbing like Jay-Z leaving the stage.

The thing is — we didn't think about what it was doing to us back then, nor did we really care. We were too busy living day to day, too focused on staying alive to worry about the long-term effects. You couldn't see the poison, so it didn't seem real. But it was in everything — the smoke, the dirt, the water that our purification specialists supposedly filtered from the Euphrates River. We were marinating in it.

One morning, I woke up and blew my nose into a tissue — it came out black. I stared at it for a second, shrugged, and went to breakfast. That was life. Nothing we could do to escape it.

The Jamaican once said, "Ya know, mon, when we go home, we all gonna glow in the dark." "Good," I said. "At least we'll save on flashlight batteries and not have to beg supply."

Even the Chaplain started joking about it. "Boys," he said once, "you may not be clean in spirit, but you're definitely purified by fire." We laughed — because if we didn't, we'd have to admit it scared the hell out of us.

Reflection

Years later, I still remember that smell. It's burned into my memory like a brand — burnt fantasies, death, and denial. The doctors back home asked if I ever worked near burn pits. I just laughed. "Worked near one? I lived in it." They didn't find it funny, but it was the truth.

Looking back, that was our real baptism — not the one in church, but the one where you learned to stop questioning why everything around you was killing you slowly. You learned to laugh, breathe, and keep going anyway. Out there, faith wasn't about God. It was about getting through another day of breathing smoke and still finding something to smile about. That was our religion — survival through sarcasm.

Army Recipe: Salisbury Steak — Burn Pit Edition
Source: TM 10-412 (Modified for Extreme Toxins and
Limited Patience)

Yield: Enough for 60 soldiers and one platoon sergeant
who swears it "tastes like Mom's."

Ingredients:
- 25 lbs mystery ground beef (or whatever came in the
  last supply drop)
- 2 lbs dehydrated onions
- 5 cups breadcrumbs (MRE cracker crumbs count)
- 1 gallon brown gravy mix (scorched chicken gravy
  optional)
- Salt, pepper, and burn pit residue for seasoning

Directions:
1. Mix ground beef, onions, and breadcrumbs in a large
   pan until your forearms cramp and you look like
   you're throwing up gang signs.
2. Shape into patties — or whatever resembles a "steak."
3. Sear on a griddle next to the burn pit for that authentic
   smoky flavor.
4. Pour on gravy until it looks edible.
5. Serve hot, with a side of secondhand smoke and mild
   regret.

Sergeant P's Note:
If your Salisbury steak tastes like a tire fire, you nailed it
— you Congratulations, you just got your daily dose of
vitamins: lead, lithium, and lube.

Chapter Ten: The Shit Talk Olympics

There are two universal truths in the military — one, you will never shit in peace, and two, if you show even the smallest weakness, and the boys will verbally destroy you like wolves on raw meat. I've been on both the giving and the receiving end of both. Welcome to The Shit Talk Olympics — where humiliation was a sport, insults were currency, and the only rule was: don't get butt hurt.

We didn't have much out there — long before cell phones were standard, no computers, no clubs, no women — so the only entertainment we had was tearing each other apart for sport. And damn, we were good at it. You could wake up, stub your toe on the corner of your cot in the dark, and someone would yell, "Peltier's crying again, somebody get him a tampon and an emotional support camel!" The laughter never stopped. If someone tripped, farted, got a Dear John letter, or even sneezed funny — it was open game for anyone.

The Jamaican was the undisputed champ. His accent made everything funnier. "Ya see, mon," he'd say, "if a girl was built like a rucksack full of rocks, I'd still take 'er for a hike." He had no filter, no mercy, and zero shame. One night, after chow, he roasted the Platoon Sergeant so hard the man just stood there, mouth open, shaking his head. "I swear to God, Private, if you weren't so damn funny and I needed you to cook breakfast, I'd smoke your ass into next Tuesday." "Ya can try, Sarge," he said, smiling. "But ya can't smoke a rock, you can only heat it up."

The Scorpion vs. Lizard Death Match
When the roasting wasn't enough, we found other ways to kill time — and sometimes creatures, as cruel as it may seem. It started as a joke. We were on guard duty one night, bored out of our minds, when one of the guys brought a tiny desert lizard in one hand and a scorpion trapped in an old water bottle in the other.

"Who you got, boys? Five-to-one odds on Team Lizard!" "Hell no," I said. "That scorpion's built like Mike Tyson with an ass needle."

We cleared a little circle in the dirt, dropped them both in, and watched nature go full UFC. The lizard darted, the scorpion lunged, and we were hollering like kids at a cage match. One guy, "Baby Boy," even held up an MRE spoon like it was a microphone, giving commentary. The scorpion nailed the lizard in the face. Game over. The Jamaican yelled, "Ya see, mon, small sting, big win!" and threw down a half-eaten melted Snickers bar from a care package as the prize. We were insane — laughing our asses off, watching tiny desert creatures fight to the death while we stood guard in the heat. And honestly? It was the most entertainment we'd had all week.

The Prank Gone Wrong

Not all the fun was harmless. Some of it crossed the line — and I was usually involved. One night, after a long shift, we decided to prank a fellow cook. He was always talking tough, walking around shirtless like he was auditioning for a WWE episode. We waited until he passed out, then carefully dropped a live lizard into one of his boots. Next morning, he slid his dirty three-week old unwashed foot in, and suddenly he was dancing like he was auditioning for Breakin'. Boots flying, screaming, "MOTHERFUCKER! SOMETHING'S IN THERE!" The lizard was probably more scared than he was, but it ran up his leg and scared him good. We all got smoked for it — pushups, sprints, burpees — but it was worth every second. The

Platoon Sergeant said, "Y'all are out here treating Iraq like its Discovery Channel After Dark!" "Roger that, Sergeant," I said between pushups. "Welcome to the desert zoo."

The MRE Bomb Brigade

Guard duty at night had its own brand of boredom. When you've stared at the same patch of desert for six hours straight, you start getting creative — and stupid. That's how we invented MRE bombs. We'd take the little chemical heater from an MRE, dump water inside the sealed bag, shake it, and seal it tight. Thirty seconds later — BOOM! It'd explode like a mini grenade. We'd toss them behind the tents, in porta-shitters, anywhere. The sound would echo across the compound like a thunderclap.

One night, the whole compound was sound asleep. It was dark and silent. I don't really know what possessed me to make an MRE bomb at that time, but once it was created, there was no turning back. This one had two chemical heaters and a liter water bottle instead of a 16 oz. I completed the concoction and threw it over the berm and waited. Nothing. It had swelled to the size of a gallon jug. I had to get it to go off and just get it over with. The other soldier from Virginia and I grabbed a few rocks, stepped back about thirty feet and started tossing them at it to trigger it. All of a sudden — BOOOOOOM! I should have been wearing a Depends because it literally scared the shit out of me. The rumble shook the area and echoed through the silent night. I knew for sure the First Sergeant was coming out and we were going to get hell. We were dying laughing and terrified at the same time. A Specialist woke up and came charging at us in full battle rattle. He told us to chill out before First Sergeant came. Next night, I woke up to find my entire bed frame zip-tied to a tent pole. Took me an hour to get down. Touché.

Reflection

That's the thing about war — the line between sanity and insanity gets blurry. We didn't prank each other because we hated each other. We did it because if we stopped laughing, we'd start thinking — and thinking was dangerous. In a world that didn't make sense, laughter was the only weapon we had left. It didn't matter if we were roasting each other, gambling on lizard fights, or making MRE explosives at 0300 — those moments kept us human. The jokes, the pranks, the insults — they were our armor against the things we couldn't control.

Army Recipe: Field Burritos — Guaranteed to End Friendships
Source: TM 10-412 (Modified for Midnight Mischief)
Yield: Enough for 8 hungry bastards and 1 explosive fart symphony (also known as the Blue Falcon).

Ingredients:
- 8 MRE tortillas (or whatever isn't moldy)
- 4 MRE chili mac packets
- 4 MRE cheese spreads (optional but mandatory for regret)
- 2 MRE jalapeño sauces
- Crushed crackers for crunch
- A sprinkle of crushed TUMS for bravery

Directions:
1. Mix chili mac, cheese, and jalapeño sauce in a canteen cup until it looks like a crime scene.
2. Spread the goo inside an MRE tortilla, add crushed crackers for texture. Roll tight.
3. Warm it on the griddle for 1-2 minutes or until it smells like poor choices.
4. Serve immediately before the next mortar attack.

Sergeant P's Note:
If your burrito doesn't make your squad clear the tent, it's not authentic. Legend of the Blue Falcon Burrito — once you eat it, you are sure to fuck over your battle buddies. Eat fast, laugh hard, and never trust a fart in a combat zone.

Chapter Eleven: Patrol Paratroopers

"Luck was our religion." That saying floated around in the back of my head every time we rolled out.

Convoy day started the same way every time: engines coughing to life like a choir of angry old men, coffee that wasn't shocked properly in the field, and that tight knot behind the ribs — not from grits, but from the feeling that someone had planted a fist there and refused to let go. We'd stack pallets of water, secure the utensils, strap down the Mer-mite containers, and then wait for the lead vehicle to bark the order: "Move out."

Outside the wire, the world changed. Inside the wire you had lines, people, rules — for some reason, saluting officers. Outside — nothing but sky and sand and the kind of emptiness that makes your thoughts loud. Convoys were strange theater: tanks lumbering like the footsteps of a tyrannosaurus rex, helicopters humming overhead like angry pterodactyls, Humvees doing laps that would make your grandmother nervous, and soft skin trucks stuffed with everything from fresh water to boxes of shrink-wrapped candy someone swore came from local village kids with names that probably weren't real.

You learned the rituals quick. Check your wheel wells.

Check your comms. Check the guy next to you to make sure he looked human, not some statue of exhaustion. Keep your head on a swivel, keep your plates facing forward, and for God's sake, don't be that idiot who laughs too loud. The desert has ears.

There was a moment — one that turned into the story we told for weeks — when the kid from Virginia let loose a .50-cal round while we were lining up to roll. It was one of those sounds that wakes the dead — a single crack that echoed like the butt cheeks of a stripper slapping the stage after sliding down the pole. The kid stood there like he'd swallowed a razorblade; his tan-brown face went white as the sand. We all saw it — muzzle flash, recoil, the panicked look — but we weren't idiots. Accidental discharges happen. In war, paperwork can be worse than punishment, so I picked up the shell casing and put it in my pocket. "Must've been a blown tire," the Sergeant said into the radio, and every one of us nodded like we'd just seen a tire explode in slow motion.

Sometimes, you save a man's ass with a story. We did it all the time. Keep the record clean, keep the boy sane, and let the rumor become fact. We kept everything in house when we could. Once we were moving, the convoy became a living thing. A rattling serpent with a bad attitude. We'd crawl along roads that passed through towns that looked like they'd been drawn by someone who hated straight lines: piles of rubble, blown-out cars, family compounds turned into fortresses. We'd see kids watching from rooftops, old men sitting on plastic chairs like judges on a panel, goats casually grazing through what used to be a front yard of some sort. Helicopters would sweep overhead, throwing shadows across the asphalt like reminder that we weren't alone.

Mount training kicked in more times than I can count. We practiced door-kicks and room-clears until it felt like muscle memory could override fear. When the call came — clear the house, take detainees, search for weapons — that training turned from motion to mission. You'd belt up, catch the glance of the guy to your left, and for one second everyone moved to the same heartbeat.

We'd kick in doors in the dark, flashbangs scorching the night, and move through rooms that smelled like charcoal and prayer. Sometimes the locals were scared stupid. Sometimes they fought back. Most of the time, we'd find weapons — AKs stacked like cordwood, RPGs that looked like they forgot they were illegal, boxes of ammo, grenades with mystery labels. It was like collecting forbidden toys for a war we didn't ask for. We'd form a line, pass them down, and load up the back of a 5-ton truck.

We'd zip-tie hands — quick, efficient, the modern handcuffs of war — and for a while we did the thing that made us feel both safe and cruel: sandbag heads. Slip a sandbag over someone's head so they couldn't see, couldn't spit, couldn't curse. One dumbass from Newport News thought he was hilarious and drew faces on the outside — big marker smiles turned into cartoon tears — and called them "sad doggin'." We cracked up until it wasn't funny anymore.

At the time, though, humor made everything easier to swallow.

Inside the homes we searched, makeshift sandboxes would be made in the corners. You'd think toys didn't matter here, but what we found in them did. Bottle-cap maps, little pellets arranged like roadways, and chalk lines with X's pointing to things we recognized as targets — not to us, but to someone's idea of a fight. Those little setups mocked you; the bottle caps were real instructions in miniature. We'd bag that stuff up, exchange glances, and wonder about the world we'd walked into.

Once we hit a checkpoint that had been promised clear, a kid would dart out with a bicycle and everything would go sideways. Or a dump truck — just a regular old dump truck — would barrel across an intersection too wide, and someone would swear it was intentional. Convoy life is a forever game of assuming everything is suspicious until proven otherwise. That's how you survive — and how you stay scared.

We seized stacks of weapons like Christmas came early. Cases of ammo, rockets, and ordnance with "Do Not Use" stickers in three languages. We counted them like farmers counting crops: this many that many, then loaded them into crates that smelled like iron and promises.

Still, humor ran through the chaos like a vein. You'd trade stories, take jabs at each other, and make bets on who'd fart next in the truck — or who would shit themselves and have to jump over the side to finish the job. You'd joke about the brass living like princes in palaces while we hauled their comforts down a dirt road. You'd steal a water from the cooler and offer it up like sacrament to a tired gunner, and he'd nod like you'd just handed him a piece of home.

There were softer moments too — the ones you didn't put on the compilation video. The soldier who shared his last peanut butter packet with a kid who'd given him a toothless grin. The look the Platoon Sergeant gave when he realized a detainee we'd cuffed was shaking like a leaf and truly terrified. You learned to make quick judgments, to be stern but humane when you could, because none of us wanted to be the monster in someone else's story.

One convoy, we rolled back through a village where locals had set out a teapot and a plate of dates and bread as an offering. We paused — brief as a breath — and took the cups like a fragile truce. For a second, the desert seemed less hostile. Then the radios crackled and we moved on.

By the time we made it back to base, the trucks were dented, the drivers exhausted, and the cooks — us — ready to reload the line. You'd drop your pack, sponge the dust off your face, and stare at the sky like you were checking for permission to be alive. Convoy life is a ledger: miles in, miles out, and a running balance of near misses. Every trip felt like flipping a coin — heads you come home, tails you don't.

We did things to tip the scales: checklists, rituals, prayers, black humor. We told the same stories until they wore down soft at the edges. They became legend. By the time we left theater, all the cooks had received Driver's Badges for driving over 100 hours in a combat zone.

And beneath all of it, there was truth: the work mattered. We weren't glamorous. We were the bloodline of the operation — the ones who kept food, water, and life moving through a country that otherwise ate people for breakfast. We got praised in passing, sometimes cursed at, but mostly ignored until something broke or ran out. Then we were essential.

Let's not forget being woken up out of sleep because we held the keys to the food containers. Soldiers would come begging for a case of energy drinks or Clif Bars before missions, sometimes offering supply materials as barter. If you were lucky, you'd make a deal with comms and sneak a two-minute call on the satellite phone.

On the long haul back one evening, as the sun sank into a smear of dust and helicopters lazily traced the horizon, I looked at the guys around me. Tired faces, cigarette stubs, a kid from

Virginia fidgeting with his radio, the Newport News joker cleaning his boots for no reason, the Jamaican talking bullshit under his breath. We'd all seen too much, done too much, and somehow laughed through more than I thought possible. We kept moving. We kept delivering. Luck was, honestly, our religion. We prayed to it, cursed it, and sometimes we bought it a drink.

Army Recipe: Chicken Cordon (Blew Up)
Source: TM 10-412 (Modified for Rough Roads and
Poor Decisions)
Yield: Enough for a convoy crew of 12 who've done too
many miles and have questionable taste.

Ingredients:
- 4 large frozen chicken breasts (or what passed for
  chicken in the supply truck)
- 1 bag instant gravy or tin of mystery sauce
- 1 sleeve saltine or MRE crackers, crushed
- 1 pack powdered cheese spread
- 1 can mixed vegetables (drained)
- 1 box instant mashed potatoes (to serve) - Salt, pepper,
  and duct tape (for moral support)

Directions:
1. Thaw chicken as best you can — sunlight under a tarp
   works if you're desperate.
2. Flatten breasts with the butt of your M4 or the edge of
   a mess kit.
3. Wrap a smear of cheese spread inside, fold, and press
   crushed crackers into the outside like breading.
4. Heat up the Garrison griddle if available. Add oil if
   you have it. Sear each side until it looks cooked
   enough to stop being questioned.
5. Pour instant gravy over chicken and simmer until
   everyone agrees it smells tolerable or until it blows up.
6. Add canned veggies to the gravy for color and dignity.

7. Make instant mashed potatoes according to package

— water boiled in an MRE bag if needed — and serve

a heaping scoop under the chicken. Sergeant P's Note:

If the chicken still clucks, you did it wrong. If it's dry, pass it around and pretend it's a protein bar. If someone complains, tell them to suck it up and make their own next time.

Chapter Twelve: Six More Months of Sand and Sanity

When they told us our deployment was extended six months, the air went still — like even the desert didn't want to hear that bullshit. You could've heard a pin drop inside the chow tent. A few trays clattered. Somebody cursed under their breath. The rest of us just sat there, silent, pretending it didn't sting like a knife to the gut.

Six more months of sand. Six more months of dust storms, burn pits, diesel breath, and bullshit. We were supposed to be heading home. We'd already made plans, booked imaginary flights, promised barbecues, weddings, and cold beers that now felt like fairy tales.

The Platoon Sergeant came in, arms crossed, jaw tight. "I know it's rough, boys. We don't control the timeline. We control our response." He always found the positive things to say to keep morale up, but we weren't trying to hear it.

The Jamaican leaned back in his chair, grinning through the frustration. "Ya mon, my response is tears, then laughter, then maybe a nap. Maybe I wake up and I'm at Fort Bragg with a big booty chick."

Someone yelled from the back, "Six more months of this shit?! I'm filing for emotional bankruptcy!"

We all laughed — the kind of laughter that wasn't joy, just survival. What else can you do? Morale dropped faster than a scud missile. Guys stopped shaving. Some stopped talking. A few stopped caring. You could see it in their eyes — that glaze, that look of men who had seen too much, felt too little, and just wanted the clock to run out.

But then, like some cosmic joke, MWR showed up.

MWR: Morale, Welfare, and "Rationed Hope"
At first, it was just rumors. Someone said they were setting up an MWR room with computers and internet. Actual internet. Not that lagging radio email crap. Then we saw it — a big semi-truck rolling in with an "MWR" sign like it was the second coming of Jesus himself.

Inside were six computers. Six. For hundreds of soldiers. Thirty minutes per person, and you had to sign your name on a list that was longer than the Bible. But it didn't matter — this was gold. Hope. Connection.

Every soldier became a desperate little nerd, creating email accounts, writing to family, pretending like they'd forgotten how to type. And then… someone discovered Hot or Not.

I shit you not — that stupid website where you could upload your picture and strangers rated you on a scale from 1 to 10. Suddenly, it became the hottest operation in Iraq. Guys were shaving, flexing, borrowing each other's Oakley's, taking pictures on disposable cameras, sending them home and having the photos sent back, scanning and uploading them to get "rated."

"Bro, I got an 8.6!"
"Bullshit, you only got that 'cause you used that photo with your rifle!"
"I'm a solid 8 with body armor on, 6 without it."

The Jamaican was the best. He uploaded a picture in his PT shorts and flip-flops, holding a tray of eggs and showing off his ashy knees. "Dis da look of a real man, mon. Strong, sweaty, and smellin' like survival." He scored a 9.1 — and wouldn't shut up about it for weeks.

For the first time in months, there was laughter again — real laughter. The sticky FHM and Maxim magazines that had been our "companions" since Kuwait suddenly took a back seat. Now the guys were typing flirty messages to random women back home, convincing themselves someone was waiting for them. Hope, even if fake, kept us breathing.

The "Movie Room"
When morale's low, soldiers turn into engineers. We found an old bombed-out barracks near the far end of the compound and turned it into our MWR annex. Someone blacked out the windows with duct-taped ponchos, a few of us dragged in some busted leather couches from the trash pit, and another guy rigged a microwave and a borrowed air conditioner from supply.

We called it "Hollywood." One power strip, one DVD player, and a rotation of bootleg movies bought off Iraqi vendors: Bad Boys II, Friday After Next, 8 Mile, and some sketchy disc labeled "THE FAST AND THE FURIOUS UNCUT" (which was not the movie we expected).

It was heaven. The smell inside was a mix of popcorn, sweat, and melted electrical cords, but nobody cared. For two hours, we weren't soldiers — we were just people again. Guys would trade guard shifts to get a seat. Some slept through the whole film just to feel cool air and pretend they were back in America. I was guilty of that more than once.

Showers and Small Miracles

Then came the real miracle — running water. Actual showers. After months of baby wipes and bottles of tepid water dumped over our heads from a water jug, someone finally connected pipes to a portable water tank. You'd think they built the damn Bellagio.

We had four shower stalls. The first time I turned that nozzle and felt actual water — not scorching, not freezing — hit my back, I almost cried. The dirt came off in layers. We all looked at each other, laughing like idiots. The Jamaican raised his hands and yelled, "its rainin' salvation, mon!"

It was the smallest thing — a shower, a movie, a half-hour on the computer — but it changed everything. For a little while, Iraq didn't feel like the end of the world.

Reflection

Six more months. That's what they gave us — more sand, more sweat, more survival. But those little things — Hot or Not, bad movies, the sound of running water — they became our lifelines. They reminded us that no matter how long we were stuck there, we were still human. We weren't fighting for medals. We were fighting for moments — thirty minutes of Wi-Fi, a cold drink, a laugh with a stranger who might someday remember our name.

Hope didn't come from orders or flags. It came from a flickering movie screen, a busted AC, and a profile that scored higher than expected. We didn't need home — just the idea of it.

Army Recipe: Chicken Tetrazzini — Extended Deployment Edition

Source: TM 10-412 (Modified for Desperate Hope and Half-Rations)

Yield: Enough for 50 soldiers clinging to sanity and sarcasm from the 6-month extension.

Ingredients:
- 10 lbs shredded mystery chicken (preferably not from last week's chow)
- 3 lbs spaghetti noodles (broken from poor supply handling)
- 2 cans cream of mushroom soup (expired but still "good to go")
- 1 quart evaporated milk
- 1 lb fake parmesan powder
- Salt, pepper, and crushed dreams to taste
- Optional: a handful of crushed crackers for "texture"

Directions:
1. Boil noodles in the least dusty water you can find. Drain while praying no scorpion lands inside.

2. Mix chicken, soup, and milk into one big pot. Stir until it looks like slop and smells like college dorm food.
3. Add noodles and stir again. Sprinkle parmesan and despair on top.
4. Bake in the MKT until morale slightly rises.

Sergeant P's Note:

If it doesn't burn the roof of your mouth off, you didn't cook it long enough. If it sticks to the roof of your mouth, congratulations — you just made it authentic. Serve with a side of gallows humor and Hot or Not ratings.

When I think back to Iraq in 2003, there's one truth that always rises above the noise: I was a Red Falcon — 1st Battalion, 325th Airborne Infantry Regiment. And being part of 1/325 AIR in that first year of the war wasn't just a unit assignment. It was an identity. A bond. A set of scars — inside and out — that we earned under a sun that baked the soul and a sky that never stayed quiet for long.

We weren't the quiet type. We weren't the "wait and see" type. We were the Red Falcons — the ones you sent forward when things got messy and dangerous and unpredictable. If someone said the roads were hot, or the locals restless, or the Fedayeen stubborn, we were already grabbing our gear before the sentence ended. We didn't walk into Iraq. We stormed into it. And every day we were there, the country reminded us we weren't on friendly turf.

People talk about 2003 like it was some clean sweep — tanks rolling fast, flags waving, everyone cheering. That wasn't our reality. Our reality was ambushes, RPGs skipping across the sky, convoys with no armor, IEDs buried in trash, and enemies who didn't fight in straight lines. It was long nights, longer mornings, and days that bled into each other until sleep felt like a fairytale. But we were paratroopers.

We adapted. We overcame. We kept pushing. We're Red Falcons. We don't break. We don't back down. We go All the Way.

About the Author

Basic Training & Airborne Photo — 2002

Nate Peltier served with the 82nd Airborne Division from
2002 to 2006 as an Army cook and paratrooper.
Born and raised in Muskegon, Michigan, he learned the craft
of food from his father, a chef, and the discipline of
leadership from his grandfather, a World War II veteran and
retired Colonel.
Now based in Kansas City, Missouri, Nate continues to
advocate for veteran mental health and post-service purpose
through the Veteran Lifeline App.
Contact: nathan.peltier1983@gmail.com

Author's Note

I didn't set out to become a soldier or a cook in a war zone — I was born into both worlds.

My father was a chef, the kind of man who could turn a handful of ingredients into a memory. My grandfather was a soldier in World War II and a retired Colonel, a leader who carried discipline in his bones and stories in his silence. Between the two of them, I learned early that food and service were the same thing — one fed the body, the other fed the soul.

When I joined the Army, I thought I'd just be flipping eggs and peeling potatoes. Instead, the 82nd Airborne Division taught me what it meant to lead under fire, laugh through fear, and keep moving no matter what burned around me. The kitchen became my battlefield; the spatula was my weapon.

This book isn't just about food. It's about the people who shared it — the ones who kept me sane, kept me humble, and reminded me that even in war, there's always room for flavor.